Eva V. Carlin

A Berkeley Year

A Sheaf of Nature Essays

Eva V. Carlin

A Berkeley Year
A Sheaf of Nature Essays

ISBN/EAN: 9783337025472

Printed in Europe, USA, Canada, Australia, Japan

Cover: Foto ©berggeist007 / pixelio.de

More available books at **www.hansebooks.com**

A Memory of Berkeley   By *William Keith*

A Berkeley Year

*A SHEAF
OF NATURE ESSAYS*

Edited by Eva V. Carlin

Published by the
Women's Auxiliary of the First Unitarian Church
of Berkeley, California

1898

A Berkeley Year

Decorated by
LOUISE M. KEELER

FROM GENESIS TO REVELATION

For the land is a land of hills and valleys; and the mountains shall bring peace to the people.

A bird of the air shall carry the voice, and that which hath wings shall tell the matter.

Consider the lilies of the field, how they grow; they toil not, neither do they spin: and yet Solomon in all his glory was not arrayed like one of these.

Wisdom hath builded here her house; she hath hewn out her seven pillars. She is a tree of life to them that lay hold upon her: and happy is the man that retaineth her.

It is a good thing to call to remembrance the former times, to remember all the way the Lord, their God hath led the people; when they were but a few men in number; yea, very few, and strangers in the land.

We have also a sure word of prophecy. Ye shall run and not be weary; ye shall go out with joy, and be led forth with peace; for the eyes of the Lord are always upon the land, from the beginning of the year even unto the end of the year.

CONTENTS

From Genesis to Revelation V

The Making of the Berkeley Hills 1
    Joseph Le Conte

They Looked Through the Golden Gate . 9
    William Carey Jones

Lang Syne 19
    Edward B. Payne

Joy of the Morning 27
    Edwin Markham

A Glimpse of the Birds of Berkeley . . . 31
    Charles A. Keeler

Walks About Berkeley 41
    Cornelius Beach Bradley

The Trees of Berkeley 49
    Edward L. Greene

On Berkeley Hills 55
    Adeline Knapp

The Love of Life 59
    Willis L. Jepson

A Berkeley Bird and Wild-Flower Calendar . . . 65
    Compiled by Eva V. Carlin and Hannah P. Stearns

*The Making of
The Berkeley Hills*

The Making of the Berkeley Hills

AMONG the many phases of out-door Berkeley, I am asked to give a brief account of that one which interests me most. Some, doubtless, would talk of the beautiful flowers which mantle the hills like an exquisitely varied carpet; some of birds, their habits, their color, their song; some would talk of the early history of Berkeley and would give reminiscences of the Golden Age of youthful Berkeley. But underlying all these, and forming the condition of their existence—without which there never would have been any Berkeley—are the Hills with their rounded and infinitely varied forms, their noble outlook over fertile plain and glistening Bay shut in beyond by glorious mountain ranges through which the Golden Gate opens out on the boundless Pacific. It was *this* that decided the choice of the site of the University, and determined the existence of Berkeley.

The Making of the Berkeley Hills

I have thus given in few words the prominent geographical features of Berkeley. But how came they to be what they are? How were they made and when? These, our beloved Berkeley Hills, were born of the Pacific Ocean about the end of the Miocene or mid-tertiary times. They took on a vigorous second growth about the end of the Pliocene epoch. Now, I well know that these terms convey little meaning to most people. Such persons will immediately ask, "How long ago was this? How many years?" I frankly confess I do not know, but I am sure it is at least a million years and perhaps much more. The geologist, you know, has unlimited credit in the Bank of Time, and he is not sparing of his drafts, as no one is likely to dishonor them.

As soon as these Hills raised their heads above the ocean, the sculpturing agencies of sun and air, of rain and rivers commenced their work of modeling them into forms of beauty. Slowly but steadily, unhasting yet unresting, the sculpturing has gone on from that time till now. The final results are the exquisitely modeled forms, so familiar, and yet so charming.

These Hills, therefore, like all mountains, were formed by *upheaval*, or by igneous forces at the time mentioned; but all the details of their scenery—every peak or rounded knob, every deep cañon or gentle swale, is the result of subsequent sculpturing by water. If the greater masses were determined by *interior* forces, all the lesser outlines—all that constitutes scenery—were due to *exterior* forces. If the one kind of force *rough-hewed*, the other *shaped* into forms of beauty.

In those golden miocene days, with their abundant rain, their warm climate and luxuriant forest-vegetation, life was even more abundant than now. The sea swarmed with animals of many kinds, but nearly all different from those we now find. The remains of these are still found abundantly in the rocks, and a rich harvest rewards the geological rambler over the hills, with hammer in hand. The land, too, was overrun by beasts of many kinds characteristic of the times. Some of these extinct animals, both of sea and land, I think, we must sorely regret; for example: little, three-toed horses, much smaller than the smallest Shetland pony, roamed in herds over our new-born hills. We have not, indeed, yet found them in Berkeley rocks, but abundantly in rocks of the same age not very far away. They probably visited our hills. We cannot but regret that these pretty little horses were too early for our boys, and indeed for any boys, for man had not yet entered to take possession of his heritage. Again: *Oysters*, such as would astonish a latter-day Californian, existed in such numbers that they formed great oyster-banks. Their agglomerated shells, each shell five to six inches long, and three to four inches wide, form masses three feet thick, and extending for miles. These are found in the Berkeley Hills; but elsewhere in California, Miocene and Pliocene oysters are found, thirteen inches long, eight inches wide, and six inches thick. Alas for the degeneracy of their descendants, the modern California oyster. And yet, upon second thought, there may be nothing to regret. It may well be that in the gradual decrease in *size* the *flavor*

The Making of the Berkeley Hills has been correspondingly intensified. It may be that what was then diffused through a great mass of flesh and therefore greatly diluted, was all conserved and concentrated into the exquisite piquancy characteristic of the little California oyster of the present day. If so, we are consoled.

But the *character* of the Berkeley Hills was not yet fully formed. Still later there came hard times for Berkeley. But hard times are often necessary for the perfecting of character, and therefore we do not regret the next age. There was for Berkeley, as for other places, an Ice-age. An Arctic rigor of climate succeeded the genial warmth of Tertiary times. Our hills were completely mantled with an ice-sheet moving seaward, ploughing, raking and harrowing their surfaces; smoothing, rounding and beautifying their outlines. The materials thus gathered were mixed and kneaded and spread over the plains, enriching the soil, and preparing it for the occupancy of man—not yet come.

Last of all—last stage of this eventful history—came man. When did he come? Was there a Pliocene man, and was his skull really found in Calaveras? If any one is interested in this famous controversy, let him consult Professor Whitney on the one side, and Bret Harte on the other.

But, certainly, evidences of *Prehistoric* man are abundant all over California, and nowhere more so than in and about Berkeley. Those interested in this subject will find abundant material.

I have thus given in bare outline, the birth, growth,

and character-making of the Berkeley Hills and Plains, in preparation for the occupancy of civilized man. The work of the Geologist is done. The Historian must take it up at this point. I have laid the ground-work; others must build thereon.

The Making of the Berkeley Hills

JOSEPH LE CONTE.

They Looked Through the Golden Gate

They Looked Through the Golden Gate

BEFORE the face of the white man came and showed that nature here was to be devoted to exalted ends, the aboriginal inhabitants had dwelt for generations on the shores that front the Golden Gate. They left mementoes of themselves at the *embarcaderos* of the creeks, Temescal, Cordonices, San Pablo, in the larger and smaller "mounds," that tell by their contents of the form and style of man himself, of his utensils and his foods. They looked through the Golden Gate, but not with the keen and perfected vision that responds to high intellectual and spiritual emotions. They lived the little life of incipient humanity, their hates and loves and a vague surmise of a Great Spirit alone testifying to the potentialities of their kind.

But one day — March 27, 1772, for 'tis interesting to fix the dates of our scanty anniversaries — representing the spiritual and temporal arms of Spain, the

They Looked Through the Golden Gate

fore-leaders of the *gente de razon*, Padre Juan Crespi and Lieutenant Pedro Fages, and their dozen companions, passed along the Contra Costa shore, and looked through the Golden Gate. They knew, indeed, that they were opposite the "mouth by which the great estuaries communicate with the Ensenada de los Farallones." They had left Monterey on March 20th; on the 25th they had encamped on Alameda Creek, near the site of the later Vallejo Mill, the ruin whereof yet standeth, or the present Niles. They crossed the San Leandro and San Lorenzo creeks and reached the beautiful *encinal*—the oak-clothed peninsula of Alameda. They passed around "an estuary, which skirting the grove, extends four or five leagues inland until it heads in the sierra," and came out upon the verdant, blooming plain. But the eye, even of the *gente de razon*, was not illumined. They sought the harbor of San Francisco underneath the promontory of Point Reyes, and searching for that which was valueless, recognized not the surpassing worth of what lay at their feet. They looked through the Golden Gate in vain.

But the Franciscans were not to be daunted in their purpose of finding their patron saint's anchorage. And so now they seek it again, this time by sea, and Juan de Ayala, Lieutenant in the royal navy of Spain, in the ship *San Carlos*, on August 1, 1775, sailed through the never-before traversed waters of the Golden Gate into the hospitable harbor. The real San Francisco was illusive; this port is now thought good enough to be dedicated to the great Saint Francis.

Then came the founding of the Mission of San Jose, June 11, 1797, under the scholarly Father Lasuen. This prosperous mission and first settlement in Alameda County was from 1803 to 1833 under the charge of the famous Father Duran. Passing up and down the shore in gradually growing numbers the Spanish Californians looked through the Golden Gate. The princely San Antonio rancho, fifteen leagues in extent, was, in 1820, conferred by Governor Pablo Vincente de Sola on Don Luis Peralta. In 1843 Don Luis, in company with his four sons rode across the domain, and with eye and gesture surveyed and partitioned it into four shares. The most southerly, in the neighborhood of San Leandro, was assigned to Ygnacio; the next, proceeding north, including Alameda and Brooklyn, to Antonio Maria; the third, covering the Encinal de Temescal, or Oakland, to Vincente, and the northernmost, including the modern Berkeley, to Jose Domingo. Peraltas, Castros, and Pachecos, worthy families in the romantic background of our history, settled along the shore and looked daily through the Golden Gate. The Castro home, at the margin of Cerrito Creek, on the San Pablo highway, screened by the Alta Punta, still yields testimony to the first habitations of the *gente de razon*.

Perhaps a broadening vision was given to the mind that daily fed upon the scene around them. They had anyhow established a settlement and a place of growing allurement to American adventure, ambition and enterprise. The American came; he looked through the Golden Gate, and his soul was uplifted. Senator Ben-

They Looked Through the Golden Gate

ton had said, his mental vision discerning its true significance, "There is the East; there lies the road to India." And Fremont, standing upon the castellated crag of La Loma, eyes filled with the refulgent beauty of the scene, senses astir with emotion, and mind prescient of potentialities, looked through, as well as *named*, that "road of passage and union between two hemispheres" THE GOLDEN GATE. On his map of 1848 he wrote opposite this entrance "*Chrysopylae*, or Golden Gate," "for the same reason that the harbor of Byzantium, afterwards Constantinople, was called *Chrysoceras*, or Golden Horn."

The fifties brought American settlers, Shattuck, Blake, Hillegass, Leonard, and others, who built their homes and prepared the land for the coming army of peaceful occupants. The American tiller of the soil looked through the Golden Gate, and his own and his children's minds were made larger and happier by the aspirations and ideals it suggested.

As Fremont had looked and had beheld with all-encompassing mind the boundless resources and possibilities springing from nature and from man's puissant hand, so now looked Henry Durant, controlled by one dominating thought. "He had set out to seek a place where learning might find a peaceful home on our Pacific shore." "And he had come to the spot, where," narrates the brilliant Felton, "rising calmly from the sunlit bay, the soft green slope ascended, gently at first, and then more abruptly, till it became a rugged storm-worn mountain and then disappeared in the sky. As he gazed upon the glowing landscape he knew he had found it. He had found what he sought through life. Not alone the glory of the material landscape drew from him the cry, 'Eureka, I have found it!' Before him, on that beautiful spring morning, other scenes, invisible save to him, passed before his mental vision. On the hill that looks out through the Golden Gate he saw the stately edifice opening wide its gates to all, the rich and the poor, the woman and the man; the spacious library loomed up before him, with its well-filled shelves, bringing together in ennobling communion the souls of the great and good of past ages with the souls of the young, fresh starters in the onward march of progress. In its peaceful walls those who had made a new goal for progress were urging on their descendants to begin where their career had ended, and to recognize no good as final save that which ends in perfect and entire knowledge. And before him in long procession the shadowy forms defiled of those to come. Standing on the heights of Berkeley he bade the distant generations 'Hail!' and saw them rising, 'demanding life impatient for the skies' from what were then fresh, unbounded wildernesses on the shore of the great tranquil sea.

"He welcomed them to the treasures of science and the delight of learning, to the immeasurable good of rational existence, the immortal hopes of Christianity, the light of everlasting truth.

"And so, hero and sage, the memory of whose friendship raises me in my own esteem, I love to think of thee. I love to think of thee thus standing on the

They Looked Through the Golden Gate

heights of Berkeley, with the strong emotion lighting thy features and the cry 'Eureka!' on thy lips, as thy gaze goes through the Golden Gate to the broad Pacific Ocean beyond."

On March 1, 1858, the site was made the permanent location for the College of California, and on April 16, 1860, it was dedicated with formal ceremony and a prayer that it might be "a blessing to the youth of this State, and a center of usefulness in all this part of the world."

Frederick Billings, name of honored power in the community, had looked, as one of this dedicatory company, through the Golden Gate. His was the inspired function to name the intellectual seat that lay facing Fremont's *Chrysopylae*. The good Bishop of Cloyne, the imperishable philosopher, who had longed for a spot "so placed geographically as to be fitted to spread religion and learning in a spiritual commerce over the western regions of the world," gave the note to Billings' inspiration, who christened the spot that looks eternally through the Golden Gate BERKELEY.

By and by others' steps were led to Berkeley, agents of the State and those who not agents were yet lovers of California. Founders of the private college had thought that the mind of youth would be broadened when their imagination might be nourished by soaring afar upon the boundless ocean. And now the planters of the State's University saw the wisdom of those who had chosen Berkeley as education's home, and accepting the gracious gift of private effort, made that the State's intellectual

They Looked Through the Golden Gate

center. And here generations of California's flower of manhood and womanhood have looked through the Golden Gate of ever-broadening insight. Dead and useless is the soul of youth or man, of student or professor, that has not daily, by nature's presence about him, felt his spirit lifted ever to higher things. An education of priceless worth is born within the mind that rightly combines in intimate development the intellectual treasures gathered in academic halls and the golden impressions that nature here unceasingly lends.

Yet once again a prescient eye looks through the Golden Gate. A home of refined and splendid architecture is to be built for a University of worthy achievement and yet richer, nobler possibilities. The world's best genius is invoked to match nature's rarest creation with art's choicest work. The mind of Phebe Apperson Hearst had looked through the Golden Gate.

WM. CAREY JONES.

Lang Syne

Lang Syne

BIRTHS and Beginnings! — the world will never weary of tracing them, that it may say, "Behold here is the seed, the plantation, from which this vital growth sprang." Especially so if myth and legend have gathered about the genesis of a man or a community, so that origins are obscured in the tinted mists of a far horizon. Ages hence some historian will curiously unwrap the dreamfolds in which Berkeley's earliest records will then be involved, and the local traditions will have antiquarian corners assigned to them in the libraries of Town and University. That this is not yet, Berkeley cannot reasonably be reproached. It got itself into human time as early as it could, and we must wait patiently until the dust has gathered on the vestiges of its origin and made them relics of antiquity.

Time, however, has wrought for us here already an ample perspective for the pictures of Reminiscence. Inasmuch as we can

Lang Syne but glance hastily at a few of these, we will not look back too far; let it be, say, to the first five years of the quarter century that ends with this year of Ninety-eight. Those who dwelt here then should be pardoned if they venture to speak of that period as "the good old times." It was the bucolic age of Berkeley, which was then, for the most part, about as God and Nature and the ploughings of a few ranchmen had made it. To be sure, Education, in its prime right, had secured and set apart for University grounds some two hundred of the most beautiful acres of Nature's wild estate. Also, about a score of dwellings were scattered here and there. But by far the larger part had the appearance of open common. The streets, (then only country roads,) were few; but numerous footpaths ran in all directions and led straight across the fields to everybody's door. There was hardly a right-angled corner to turn, in all the eastern portion of the town. Even the iron rails of the S. P. turned aside in a graceful curve to avoid the immovable cabin of Mrs. ———n. Detached patches of grain and hay ripened under the July sunshine. Everywhere else the assertive tarweed flourished, to smear with its black mucilage the trouser-leg and the trailing skirt. The summer tradewinds caught up a glory of dust into clouds that rivaled the fog. In clear and quiet weather each dwelling enjoyed an unobstructed view of the Bay, and the opening into the Pacific seemed so wide and ample that every resident, from Temescal almost to San Pablo, claimed for his own house the distinction of being "exactly opposite the Golden Gate." The hills, eastward, held out as to-day their irresistible invitation to the stroller, but wore the grace of a more perfect solitude than now. One might wander there all day and be utterly alone except for the browsing kine, the bleating sheep, and the inquisitive ground squirrel. The glistening roofs of Oakland and San Francisco appeared to be farther away than now from the lonely and rugged summit of Grizzly. Indeed, all Berkeley seemed much closer and more akin to nature than to the world of men. Alas! (though this may be lamentable to only a reminiscent mood,) that a city should have arisen here, driving back the line of Nature's outposts, and covering her simplicities under a crust of civilized improvements!

Even the University was not so imposing as to-day, and seemed to the visitor more like a pioneer home of learning than an institution of world-wide relations and reputation. No one can begrudge to education the multiplied facilities of the present time, but there was much that is memorable in the status of those early days. Characterized as it was by experimentation and the struggles incident to scanty resources and the uncertainties of popular support, it challenged the sympathetic and active interest of all lovers of liberal culture, and at all times, the little community here was a unit in championship of the University as against the outcries of prejudiced parties throughout the State.

Perhaps this committal to a common cause was what gave to the people of the place a social unity also in that period. Moreover, we were hardly many enough then for factions and cliques, and the tracing of those occultly

Lang Syne determined lines which mark off social zones and temperatures. We enjoyed that pioneer sense of a general community of interests which characterizes the early stages of every growing society. Alas! that it so invariably passes, when the tally of social units becomes the census of a multitude! How will it be, we may wonder, with the one hundred and forty-four thousand, bearing the seal of sainthood, and gathered out of the earth, according to the Apocalypse, to swell the happy population of heaven?

However it may be with the angelic multitude in the future day, it is certain that the distinctly human and earthly dwellers in Berkeley, twenty and twenty-five years ago, were disposed to a generous and genial social grace. The free sociability of that time is a happy memory. The paths joining dwelling to dwelling were the worn ways of an impartial good-neighborhood. So, also, the trails among the hills; they testified to the ramble and loiter of a chummy comradeship unchilled by hesitations. And it was even true that for a considerable time we had here but a single church, in which the variant faiths forgot their divergencies and coalesced in a unity of the spirit for the worship of the One Father. Good old times!

Some of the conspicuous figures of that earlier circle still move in the larger round of Berkeley life. They need not to be named here; they are among the specially honored citizens of our present day, or hold their places in the University faculty through the deserts of their fidelity, wisdom, and beneficent achievements. Others are now elsewhere in the world of men, putting their hands to useful task and honorable service. And yet others have "crossed the bar," and sailed forth through "Gates of Gold" to that far continent of our faith, built of "the substance of things hoped for."

May we not fittingly name two or three of these last, in token of a memory as touching them which no autumn of time will cause to fade and grow sere? Among them was C. T. H. Palmer, whose native keenness of intellect, and preeminent social geniality transmuted even a disability into a much appreciated advantage, as an ictus for his ever ready wit, or for the incisive utterance of his unfailing word of wisdom. There was Edward Rowland Sill, whom to know in intimacy was to dwell in the presence of a living poem, in which the notes of Nature, the accents of the Infinite Spirit, and the holy passions of a human soul all sang in harmony, prophesying of vital truth. There, too, was that scholar of foremost rank, the elder Le Conte. For in those days there were two to be venerated and beloved under that honored surname; although we more habitually "had reverence to them," (to adopt Mrs. Partington's felicitous misuse of a word,) by substituting those titles of special and affectionate distinction — "Professor John," and "Professor Joe." There were others also with us then—like Hamilton, who dwelt for a time among the trees on the initial lift of yonder hill—who have since joined the Choir Invisible. These are now of those "shadow men," departed out of the flesh, but living among us still through the

Lang Syne vital persistence of the spirit, and our imperishable remembrance of their words and deeds.

But now as these last lines are written the bells are ringing in an autumn day of this 1898. A glance through the open window reveals a new Berkeley, the hale and vigorous growth of a quarter century, testifying to the developing power of time, under the guidance of a dynamic idea such as Education. In this scene the vestiges of the old Berkeley are few, and some of them not easily traced. North and South Halls stand yet on their conspicuous sites, to give way eventually, no doubt, before the already invoked genius of the world, bringing in an architecture proportionate to Nature's work as here displayed. There are also yet to be seen most of the few houses of the former time; but when memory knocks at the doors it is only to be met by strange faces and new voices. The Old has had its day; the New is here, and prevails in its incontestable right. And while we cherish the reminiscent pictures of the Berkeley that was, we rejoice in the Berkeley that is and is to be.

EDWARD B. PAYNE.

*Joy of
The Morning*

I HEAR you, little bird,
    Shouting aswing above the broken wall,
    Shout louder yet: no song can tell it all.
    Sing to my soul in the deep still wood
    'Tis wonderful beyond the wildest word:
    I'd tell it, too, if I could.

Joy of the Morning

Oft when the white still dawn
Lifted the skies and pushed the hills apart,
I've felt it like a glory in my heart—
(The world's mysterious stir)
But had no throat like yours, my bird,
Nor such a listener.
                EDWIN MARKHAM.

A Glimpse of
The Birds of Berkeley

A Glimpse of the Birds of Berkeley

AS the seasons come and go, a host of birds tarry within the confines of Berkeley, some to make their nests and rear their broods, others to sojourn for but a brief interval in passing from their summer to their winter haunts, and in the joyful return of spring. They inhabit the spreading branches of the live oaks, and the open meadows are their home. They dwell in the leafy recesses of the cañons and haunt the shrubbery of our gardens.

It is impossible to understand our birds without knowing something of their surroundings — of the lovely reach of ascending plain from the bay shore to the rolling slopes of the Berkeley Hills (mountains, our eastern friends call them); of the cold, clear streams of water which have cut their way from the hill crests down into the plain, forming lovely cañons with great old live oaks in their lower and more open portions, and sweet-scented laurel or bay trees crowded into their narrower and more precipitous parts;

A Glimpse of the Birds of Berkeley of the great expanse of open hill slopes, green and tender during the months of winter rain, and soft brown and purple when the summer sun has parched the grass and flowers. These, with cultivated gardens and fields of grain, make the environment of our birds, and here they live their busy lives.

There comes a morning during the month of September when a peculiarly clear, crisp quality of the air first suggests the presence of autumn. It is something intangible, inexpressible, but to me vital and significant of change. In my morning walk I notice the first red tips upon the maple leaves, and catch the first notes of autumn birds. I hear the call of the red-breasted nuthatch, a fine, monotonous, far-away pipe, uttered in a succession of short notes, and upon looking among the live oaks, detect the little fellow hopping about upon the bark. He is a mere scrap of a bird, with a back of bluish gray and a breast of a dull, rusty-red hue, a cap of black and a white stripe over the eye—a veritable gnome of the bark, upon which he lives the year round. In its crannies he pries with his strong, sharply-pointed beak for his insect food, and in some hollow his little mate lays her eggs and rears her brood. With so many woodpecker traits he nevertheless differs widely in structure from that group, being more closely allied to the wrens and titmice. He is with us in greater or less abundance throughout the winter, and his very characteristic call may be heard from time to time both in the University Grounds and in the cañons.

With the nuthatches, come from their northern breeding places, the pileolated warblers, and other shy wood-creatures which haunt the quiet, out-of-the-way nooks, and shrink from the presence of man. The pileolated warbler is one of the loveliest, daintiest creatures that visit us. As I walk in my favorite nook in the hills, Woolsey's Cañon, to the north of the University Grounds, I see a lithe, active, alert little bird, gleaning for insects among the leaves, now high up among the branches, and again darting hither and thither downward to where the fine thread of water has formed a pool, there to bathe an instant and then, with a lightsome toss of spray flirted from its wings, to resume its quest among the bay leaves. It is a waif of gold with a crown of jet, and its song, a sweet, sudden burst of woodland music, is quite in keeping with the singer.

Let me picture my cañon in the autumn time, when the open hill-slopes are covered with tarweed and dead grass, and the country roads are deep in dust. There is a quiet, almost sacred feeling about the place, shut in by steep hill-slopes, crowded with bay trees through which the sun filters in scattered beams, and carpeted with ferns and fallen leaves. Bulrushes, with their long, graceful filaments encircling their jointed stems, spring from the tangle of shrubbery, and the broad, soft leaves of the thimbleberry, now beginning to turn brown, fill in the recesses with foliage. Great slimy, yellowish-green slugs cling to the moist rocks, and water-dogs sprawl stupidly in the pools.

A loud, ringing call sounds above as a flicker comes our way and announces his presence with an emphatic

A Glimpse of the Birds of Berkeley

ye up! He is with us all the year through, and an interesting fellow, I have found him. Not wholly a woodpecker, and yet too closely related to that family to be widely parted, he is an anomaly in the bird world. Sometimes he alights upon the ground and grubs for food like a meadow lark, while again he hops in true woodpecker fashion upon the tree trunk, pecking holes in the bark. He has the proud distinction of being the only California bird which habitually intermarries with an eastern representative of the genus—the golden-shafted flicker of the Atlantic States and the red-shafted flicker of the Pacific region intermingling in a most bewildering way, so that hybrids are almost as numerous in some sections as the pure species.

The flicker is a large, showy bird, somewhat greater than a robin in size, with a conspicuous white rump-patch, and with the shafts and inner webs of the wings and tail colored a bright scarlet. The male bird is also adorned with a streak of the same color on each side of the throat. The back is brown, closely barred with black, and the under parts are pinkish buff, marked with a large black crescentic patch on the breast and conspicuous round black dots on the lower portions of the body.

In the spring time the flickers bore a deep hole in a decayed oak limb and the mother bird lays there ten or more of the most beautiful eggs which ever gladdened a mother bird's heart, save that I fear her little home is too dark to give her so much as a peep at her treasures. They are white, with a wavy texture, like water marks in the shell, and, when fresh, beautifully flushed with pink, more delicate in color than a baby's ear. When the young brood are all hatched what a clamoring and calling there is about that hole, what an array of hungry beaks are thrust out awaiting the morsel that the busy parent carries to them! But now, in the autumn time, the family cares are ended and the flicker roams the woodland contented and well fed. Long may his piercing, buoyant call ring amid our hills, and his coat of many colors adorn our landscape!

I cannot speak of noisy birds without recalling the jays, for they are the noisiest, rollicking, happy-go-lucky fellows that make their home in our cañons. They laugh and screech by turns, they question and scold. Even when on the wing they utter a succession of loud, insistent call notes, and upon alighting, mischievously question in a shrill squeak, "*well? well?*" I am speaking of the California jay which is the common species about Berkeley,—a long, rather slender fellow, without a crest such as the blue-fronted jay of the redwoods possesses. Its back is colored blue and brownish gray, and its breast is a lighter gray, edged and faintly streaked with blue. Its manners are often quiet and dignified when sitting still and eyeing an intruder, not without a half scornful, half inquisitive glance, I fancy; but with a sudden whim it is aroused to animation, flirting its tail, bending its head on one side and suddenly fluttering away with a loud laugh.

Another of my cañon friends is the wren tit, a bird which is found only in California, and without a counterpart, so far as I am aware, the world over. It is a

A Glimpse of the Birds of Berkeley

friendy little fellow, considerably smaller than a sparrow, but with a long tail usually held erect in true wren fashion. Its plumage is soft and fluffy and its colors as sober as a monk's, brown above and below, but somewhat paler on the under portions where a tinge of cinnamon appears. The wren tit is a fearless, friendly little creature, hopping about in the tangle of blackberry vines almost within reach of my outstretched hand, but so quiet are its colors and so dense the thickets which it inhabits, that the careless eye might well overlook it. The little low chatter which it utters tells us of its presence, and if we wait quietly for a moment it may even favor us with a song. It is a simple strain, a high-pitched pipe — *tit–tit–tit–t r r r r r e!* but a sweet and characteristic note in our cañons.

As autumn moves on apace the winter birds assemble in full force. The golden-crowned sparrows come flocking from their Alaskan and British Columbian homes, and the Gambel's white-crowned sparrows from their breeding places in the mountains,—the one adorned with a crown of dull gold, black bordered, and the other with a head marked with broad stripes of black and white. Both have backs of streaked brown and gray, and breasts of buff or ash. They are among our commonest and most familiar winter residents, dwelling in our gardens as well as in the thickets among the hills, and singing even during the milder rains. The call note of both species is a lisping *tsip*, and their songs have the same quality of tone—a fine, high, long-drawn whistle. I have written down the most usual song of each species in musical form, and repeat them as follows. The golden-crowned sparrow sings:

A Glimpse of the Birds of Berkeley

The song of Gambel's sparrow is a trifle more elaborate, commencing on an upward scale, instead of the downward, as in the former case. Loud and clear comes from the rose bushes the treble whistle;:

Gambel's sparrow sings not only all day long but occasionally at night. Often upon a dark, misty night in February or March I have heard a sudden burst of bird music, and recognized the very clearly-marked strains of this bird. Coming out of the dark, damp night, so sudden and so beautiful, and followed by so perfect a calm, I know of no more impressive bird music.

When the rainy months of winter are ended and the meadow lark is sounding his loud, rich strains from the field, and the linnet is fluttering and bubbling over with song, a host of merry travelers come hurrying to our trees and gardens. The jolly little western house wren

A Glimpse of the Birds of Berkeley

bobs about in the brush, and, as the wild currant puts forth its first pink, pendulous blossoms, the beautiful little rufous humming-bird comes to dine upon them. I know not how he times his visit so closely, but certain it is that the pungent woody odor of these blossoms is inseparably linked in my mind with the fine, high, insect-like note of these pugnacious little mites in coats of shimmering fire, that come to us from Central America at the very first intimation of spring.

In April arrive the summer birds, full of the joy of the mating season. The Bullock's oriole, clad in black, orange, and gold, sings its loud, elated strain from the tree tops, the black-headed grosbeak carols in the orchard, the lovely, little, blue-backed, red-breasted lazuli bunting warbles in the shrubbery, and finally, the stately, russet-backed thrush, quiet and dignified in his coat of brown, with white, speckled breast, the most royal singer of our groves, sends forth upon the evening air such sweet organ tones that the whole night is full of melody.

I would that our birds might receive some measure of the appreciation which is due them, and that we might all turn at times trom the busy affairs of life to listen to their sweet songs and winning ways. May they ever find within the confines of Berkeley a haven of refuge from that merciless persecution which is steadily reducing their numbers. May they find here loving friends ready to champion their cause, and may they ever be considered the chief ornament of our hills and gardens!

CHARLES A. KEELER.

Walks About Berkeley

Walks About Berkeley

THE casual observer might find very little of promise in the Berkeley hills to lure him on to their exploration. Their brown slopes, wrinkled and threadbare as the sleeve of a hunter's jacket, seem to reveal to the very first glance all that they hold in store. No surprise, surely, can be waiting for one on those bare, open hillsides. The imagination pictures no secret nooks, no wooded ravines, no crag or waterfall behind the straggling screen of fern and scrub that fringes its waterways. Yet, after all, the charm of surprise is a veritable feature of the walks about Berkeley — surprise not keen and startling, to be sure, but genuine and of the quality that does not pall by frequent repetition. Thus it is that the number and variety of these rambles is a source of unending pleasure to those who have come to know them. There is a large gradation too in their extent and in the effort they require:—the quiet saunter up Strawberry Cañon in the gloaming, the long afternoon ramble

Walks About Berkeley over the hills to Orindo Park, the all-day tramp by the Fish Ranch to Redwood Cañon and Maraga Peak, or more strenuous still, the cross-country trip to Diablo. You may follow the quiet country lanes with pastures, orchards, and grain-fields dotted about here and there among the enveloping wildness. You may even find abandoned roadways leading nowhither, constructed at large expense by some one who surely was a lover of his kind, and now bequeathed to your sole use and behoof. You may thread some cool, mossy ravine where the stream runs deep in its rocky channel, under a close roof of alders and redwoods. Or you may breast the steep slope, each step opening up a wider and wider prospect, until from the east you catch the exultant flash of Sierra snows, and on the west, far beyond Golden Gate and Farallones, you gaze with awe on the immensity of the Pacific.

I do not mean to weary the reader with an itinerary of these various routes, or a tabulation of their peculiar charms. Such things are best learned when they come with the zest of discovery. To one quaint nook only would I offer to conduct my reader, and with the more reason, perhaps, because while it is easy enough of access, it seems to be very little known. The place is called Boswell's, though why so called I have never been able to guess. The name suggests human habitation at least, if not also vulgar resort and entertainment; but both suggestions are wide of the mark. Our visit shall be on some bright morning in April. We take the train to Berryman station, and zig-zagging thence northwestward, we soon are clear of the thin fringe of dwelling-houses, and out among the fields. Our course so far has been as if for Peralta Park; but instead of turning sharply down to the west at the margin of a little creek, we cross the bridge, and follow the country lane northward. When the lane also turns abruptly westward, some half-mile further on, we abandon it altogether, continuing our former direction over fields and fences, and across two little waterways. Beyond the second rivulet we reach a broad slope thickly strewn with rocks and boulders, and dotted about with low trees and shrubs. This is Boswell's.

The air all along has been full of the sounds and scents of spring: — the gurgling notes of the meadowlark, the rich smell of newly ploughed fields, the warm breath of mustard in bloom. But this untamable rock-strewn area, like the Buddhist monasteries of the far east, has become a veritable sanctuary for plants and living creatures that could not maintain themselves in the open in their unequal struggle with that fell destroyer, man. Here the wood-rat has piled undisturbed his huge shelter of sticks. The warbler and the thrush are singing from every covert. The woodpecker and the squirrel shadow you from behind tree-trunk or rock to discover your intent in trespassing thus upon their private domain; while the flycatcher flashes his defiance in your very face, if you venture too near his mate on her nest. Nor is it otherwise with the plants. Delicate species that are fast disappearing before cultivation — the blue nemophila, the shy calochortus, the bright pansy-violet—

bloom here undisturbed in all their pathetic beauty. "If God so clothe the grass of the field, which to-day is, and to-morrow is cast into the oven, shall he not much more clothe you, O ye of little faith?"

But we linger here too long upon the threshold. The tract is a considerable one, and midway there is thrust up into it from the west a sombre wedge of eucalyptus forest, contrasting strangely with the rest of the scene. For here we seem to be in a region three thousand miles away,—in a veritable bit of New England hill-pasture with its labyrinthine paths, its ever-changing short vistas, its endless series of little secluded alcoves walled about with shrubbery and carpeted with grass and flowers. The rocks too are of striking size and form, and culminate near the lower end of the tract in a bold, fantastic crag, in itself well worth the effort to visit it. But the most unlooked-for feature of the place is its air of remoteness and seclusion. Here it lies, spread out on the open hillside, in full view from bay and from town. Yet as we thread its quiet alleys, or lie dreaming in the sunshine under the lee of its rocks, we seem to have journeyed leagues from the work-a-day world we left behind us but an hour ago.

It is good to be here! And good it is also to return to the world. The joy of the scene and the season, the clearer brain and quickened pulses we shall bring back with us as we take up again the effort and struggle. And more than this we may sometimes bring from such a sanctuary,—some heavenly vision,—some far-seen glimpse of a transfigured life that may be ours,—in the

strength of which we shall go many days, even unto the *Walks*
mount of God. *About*
Berkeley

Cornelius Beach Bradley.

The Trees of Berkeley

The Trees of Berkeley

AMONG many happy remembrances of the Californian out-of-door world which some sixteen years of residence on that delightful coast have left with me, to stay while memory lasts, is that of the Berkeley landscape. And one cherishes such a mental picture as that of those massive hills, with undulating slopes and rounded summits, all verdure-clad and flowery, almost from the beginning of the year till midsummer; then for succeeding weeks as beautiful with a kind of harvest-field yellow, this deepening into brown as autumn days draw near; and always varying in their beauty with every change in the everchanging sky; beautiful under cloud, and in sunshine; beautiful in the light of early morning, in the effulgence of noonday, and at the setting of the sun.

And this fine picture of the higher hills has a rich foreground in the groves and thickets which adorn the lower slopes and thence extend to the plain below. Alders throughout the northern zone

The Trees of Berkeley follow the water courses in hilly districts, but usually as a fringe of shrubs; but here in the Berkeley cañons they are trees, and shapely ones, almost replacing the admired beeches of our Eastern States and of Europe; the beech being absent from California. And above the alders, on drier ground flourishes the California Laurel; this, in its compact habit, perennial verdure, keen fragrance of foliage, and in the beauty of its wood, having no compeer among its own kindred on our continent.

And the more humble woody and bushy growths associated along the stream-banks with the trees aforenamed, in their own way surpass them in grace and beauty. Such are the pink-flowered wild currants; and even the wild gooseberries native to these hills; and these last, though they yield but prickly and insipid fruits, more than compensate for this at flowering time by the strongly contrasted clear white and deep red or dark purple of their large almost fuchsia-like flowers; these being put forth in profusion often before the mild winter season of Berkeley is past. A few weeks later and the ceanothus bushes, masses of bloom intensely blue, are seen intermixed with the soft plume-like white panicles of the wild spiræa; the two together, or either one alone, charming every lover of the flowery out-of-door world.

The groves which formerly covered all the comparatively level country that lies along the bases of the hills, and of which considerable remnants are still to be seen, especially on the University grounds, consisted mainly of the native oak, with more or less of the Californian Buckeye, or Horsechestnut intermixed. Within the last forty years many exotic trees have been planted, either among the oaks, or in masses apart from them, where they now form separate groves. But it is interesting to note in what perfect keeping with the landscape of rounded hills above them the native oaks are, as to form and outline. For all of them, however large, present a comparatively low, broad, and evenly rounded figure, exceedingly unlike that of the oaks of other countries, and exactly harmonizing with the general outlines of the Californian coast hills whose bases they adorn. Entering under one of these oaks, the trunk is seen to be parted from near the ground into ten or more, each separate trunk extending upwards half horizontally, in such wise that the horizontal extent of the tree as a whole quite exceeds its height; and occasionally one or more of the arms of the trunk almost recline along the ground; thus affording not only a deep shade, but a resting place for the out-of-door saunterer who enters this leafy retreat. And, our oaks retain their verdure throughout the year. Without being evergreen in the strictest sense, yet, the leaves of one season remain fresh and in place all through autumn and winter, and are only ready to fall when the foliage for the new year is almost full-grown in April.

The Buckeye is also, in a smaller way, broad, rather than tall, and offers almost as deep and banyan-like a bower of shadiness in summer as the oak; and in flower, with its long spindle-like garlands of pale pinkish bloom, is one of the finest ornamental trees of which any land can boast.

The Trees of Berkeley

Somewhat later in the summer than the flowering of the Buckeye, there appear the rather dull-white clusters of the bloom of the Christmas Berry, or Californian Holly; a small tree, and evergreen; not at all conspicuous in flower, yet, in November and December days, when its ample bunches of berries have ripened to rich crimson, easily rivalling the real Holly in its beauty.

The exotic trees which have found the Californian soil and climate congenial, and which have come to form a notable element in the Berkeley landscape, are so numerous in species that one must not attempt to name half of them, where space is limited; but there are some which should not here be left without brief mention. The large Eucalypti, for example, when growing singly or in small groups among the native oaks, and towering far above them, have not only a certain combination of grace and majesty of their own, but give a variety to the landscape which is most pleasing.

And again; the Cassias, so surpassingly beautiful when, at the end of winter, they deck themselves completely in soft sprays of feathery yellow bloom—these in all their varieties, unite with lilac and laburnum, almond-tree and apple-tree, and a host of other flower-bearing tree-growths, to make the Berkeley parks and ways in spring fair and fragrant as the paths of Paradise.

EDWARD L. GREENE.

On Berkeley Hills

On Berkeley Hills

HE sun lies warm on Berkeley hills:
The long, fair slopes bend softly down
To fold in loving arms the town;
The sun-kissed uplands rise and swell,
And blue-eyed-grass and pimpernel
Dot the young meadow's velvet sheen.
The air with spring-time music thrills,
Sweet songs of birds in halls of green
    On Berkeley hills.

The sun lies warm on Berkeley hills:
The poppies gleaming orange-red
Down the broad fields their mantles spread;
Beyond the marshes glints the Bay,
Its islands lying brown and bare
Leviathan-like sunning there.
Brave ships are sailing through the gate,
The wind their spreading canvas fills—
It whispered through the trees, but late,
    On Berkeley hills.

The sun lies warm on Berkeley hills:
Across the Bay, from misty view
The City rises toward the blue;
With feet of clay, with burdened wings,
Yet pressing up to better things
From level height to level height!
Here where the hush all clamor stills
Her beauty shows, a goodly sight,
    From Berkeley hills.

On Berkeley Hills

The sun lies warm on Berkeley hills:
The wide gate beckons out to sea,
Swift birds above, poised high and free
Invite the soul to golden flight
To where there open on the sight
Large visions of that coming day
When faith that sees, when hope that wills
Shall bring man's best to dwell alway
  On Berkeley hills.

ADELINE KNAPP.

The Love of Life

MANY years ago, just as the fairy books have it, the entire Berkeley land from the summit of the hills, flecked with cloud-shadows, to the sands of the bay shore, lying naked in the sun, belonged to the wild flowers and their friends, the trees and shrubs. The right of the flowers, children of the Sun, to possess the cañons, slopes and fields, is of exceedingly ancient origin: nurtured by Mother Earth, heedful of the call of the Rain God, responding to their guardian, the Sun, they made annual proclamation of their title. Each year the wealthy Lupine family came forth to give the sign; lowly Nemophilas chose their places; Brodiaeas, purple-stalked, joined the company, while round about, leaving nowhere a vacant place, innumerable throngs of parti-colored Gilias followed in the crimson wake of Calandrinia. Hundreds of zealous retainers joined this foregathering of the inhabitants of the fields. On the remoter landscape, the Baerias filmed the ground with gold, while high on some half-inaccessible cañon wall, Godetias and Clarkias, crimson-mouthed and scarlet-lipped, stood as beautiful as victory.

The Love of Life

After the caballero, came our own house-building and pasture-inclosing people who left scarcely a " common " where the delicate " first inhabitants " might live as they had lived in the old days, but appropriated nearly every bit of meadow and hill-slope to themselves.

And still our people, not content with so much, trooped out of their houses at that season when the apple-blossom comes again on the tree, and made unceasing war on the flower people, especially on those most graceful or engaging, so that a blossom raised its head in overflow of happiness only to meet death. Those people who, perhaps being lazy, came not early enough, returned into the houses empty-handed, or pulled green branches from the trees and shrubs (because they are the friends of the flowers), stripping down the bark and leaving long gaping bleeding wounds.

Now some of the wild flowers retreated into the hills, some found half-secure hiding places in the edges of thickets, and some were never seen again. But a few others, hardy adventurers, returned each year with the passing of the winter rains. Do you not wonder that this is so? Why is it? It is because of the overmastering love of life, which is their inheritance, and the endless pains that the plant takes to secure its own safety and the safety and highest welfare of its children. Blue Dicks people the south cañon-sides, a glad company, because Blue Dick keeps most of his precious body deep in the ground and there providently stores food against blossoming and seed-making time. A handsome fellow is Blue Dick in the month of March, with his light-blue flowers hugging close together and their royal purple coats thrown half back, the whole cluster of them raised on a leafless stalk. As for the leaves, they are very long, and you will find them close to the ground.

The Yellow Violet is just such another contriving plant. The enemies of him pull him up by the roots, or think they do, not knowing, luckily, that the coral-like strands which are torn from the ground are not roots after all but only underground stems. The real roots lie very deeply buried and, so, the Yellow Violet goes bravely on flowering year after year, striving to bear seedpods that its family may increase in the land of open woods.

In April, King's Cups sprinkle the fields, the yellow flowers borne in such nest-like rosettes of leaves that some of us call them Golden Eggs! The spreading petals terminate a long thread-like tube that runs down almost into the ground where the seed-bearing part is hidden out of harm's way. What eqxuisite care is this! What bolder expression of the desire to live!

In February and April, Buttercups color the pastured hills for leagues and leagues, brilliant in the sun, appearing on the distant slopes as if painted into the very texture of the earth itself. Are you not ready to ask why grazing animals do not like Buttercup leaves and buds? The Buttercup knows why! Of this we may be sure: if ever grazing animals once found the Buttercup palatable, then there would never be a second generation of Buttercups.

Some time we shall see more of the wonderful things in Nature and so shall the wonder grow that we shall forget our primitive instincts and delight no more in the hunter's joy, the kill for the sake of the kill. Some time there will be here in Berkeley a wild-flower protection society, just as in older states, and those who have wide grounds will give the wild flowers a corner—all their own. Some time, gentle reader, the call will come down from the mountain top and you shall come up from the valley and go on a little journey over the hills on a rainy April day, the high grass wet, the west wind blowing,

The Love of Life and with new perceptions the true story of the wild flowers will be told you in every gesture of leaf and curve of bud. Doubtless the flowers are happiest when the sun shines; when their gay colors signal the passing bee or butterfly, carriers of pollen, the transfer of which, as you know, makes better seeds and seedlings and the seedlings better and larger plants. But in stormy weather, when the rain drops are falling and you can hear the sound of water running in the gulches, some of the most curious and interesting features of their lives are disclosed to even the least sympathetic observer: behold the eager attitude of their leaves stretched out for light, the way in which they keep warm, the ingenious manner in which protection is secured against rain. These are some of many things that will excite your senses, and then your responsive nature will find on every hand the choice inhabitants of the hills warm with emotions, on every side you will see the effort for self-preservation, everywhere the expression of the overmastering desire—the love of life.

WILLIS L. JEPSON.

A Berkeley Bird and Wild-Flower Calendar

A H! well I mind the calendar,
Faithful through a thousand years,
Of the painted race of flowers,
Exact to days, exact to hours,
Counted on the spacious dial
Yon broidered zodiac girds.
I know the trusty almanac
Of the punctual coming-back,
On their due days, of the birds.

Berkeley Bird and Wild-Flower Calendar

RALPH WALDO EMERSON.

I DO not want change: I want the same old and loved things, the same wild-flowers, the same trees and soft ash-green; the blackbirds, the coloured yellowhammer sing, sing, singing so long as there is light to cast a shadow on the dial, for such is the measure of his song, and I want them in the same place—let me watch the same succession year by year.

Proem: The Pageant of Summer.

RICHARD JEFFERIES.

January Birds

Townsend's Solitaire. Very rare.
Lutescent Warbler. Common resident.
Pine Finch. Occasional in flocks during winter.
California Woodpecker. Common at times in winter.
Western Golden-crowned Kinglet. Fairly common winter resident.
Ruby-crowned Kinglet. Abundant during the winter months.
Western Robin. Common at times in flocks in winter.
Western Winter Wren. Rare winter visitant.
Dwarf Hermit Thrush. Common, but shy winter resident.
Western Blue-bird. Common at times in flocks.

> It's little I can tell
> About the birds in books;
> And yet I know them well,
> By their music and their looks.
> When Spring comes down the lane,
> Her airy lovers throng
> To welcome her with song,
> And follow in her train:
> Each minstrel weaves his part
> In that wild-flowery strain,
> And I know them all again
> By their echo in my heart.
>
> HENRY VAN DYKE.

Pussy Willows.   Along creek banks.
Blue Hound's Tongue.   Thickets of the cañons.
Chickweed.   In the shade of walls and fences.
Shepherd's Purse.   Common in field and by roadside.
Flowering Currant.   In cañons and along streams.

 Thou sendest forth Thy Spirit; they are created; and Thou renewest the face of the earth.

DAVID THE PSALMIST.

Pleased Nature's heart is always young,
Her golden harp is ever strung;
Singing and playing, day to day,
She passes happy on her way.

JOHN VANCE CHENEY.

February Birds

Gambel's White-crowned Sparrow. Very abundant in flocks.
Golden-crowned Sparrow. Abundant in flocks.
Samuel's Song Sparrow. Very common resident.
Oregon Junco. Common in flocks during winter.
Townsend's Sparrow. Common, but solitary.
Oregon Towhee. Common resident of the cañons.
California Brown Towhee. Very abundant everywhere.
American Goldfinch. Locally distributed in flocks.
Evening Grosbeak. Very rare.
Cedar Bird. Occasional in flocks.

The endless, sweet reiterations of birds mean something wiser than we dream of in our lower life here.

HARRIET BEECHER STOWE.

Do you ne'er think what wondrous beings these?
Whose household words are songs in many keys,
Sweeter than instrument of man e'er caught!
Whose habitations in the tree-tops even
Are half-way houses on the road to heaven!

HENRY W. LONGFELLOW.

How fitting to have every day in a vase of water on your table, the wild-flowers of the season which are just blossoming. **February Flowers**

HENRY D. THOREAU.

Trillium.   In heavily-shaded cañons.
Wild Cucumber.   Ivy-like; over stumps and shrubs.
Indian Paint-Brush.   Rocky points of the hills.
Wood Sorrel.   In sunny, sheltered corners.
Leather Wood.   In Strawberry Cañon.
Indian Lettuce.   Shade of oaks and laurels.
Dandelion.   A bright apparition of field and meadow.

Dear common flower, that grow'st beside the way,
My childhood's earliest thoughts are linked with thee.
To the Dandelion.

JAMES RUSSELL LOWELL.

Simple and fresh and fair from winter's close emerging,
Forth from its sunny nook of shelter'd grass—innocent, golden, calm
   as the dawn,
The spring's first dandelion shows its trustful face.

WALT WHITMAN.

March Birds

Burrowing Owl. Found in the hills. Becoming scarce.
Western Screech Owl. Resident. Common.
Barn Owl. Formerly common about town. Now rare.
Western Great Horned Owl. Occasional in the woods.
Barn Swallow. Common.
Cliff Swallow. Abundant.
California Partridge (Valley Quail). Fairly abundant.
Pileolated Warbler. Solitary as a rule.
Brewer's Blackbird. Abundant in flocks.

There is something almost pathetic in the fact that the birds remain for ever the same. You grow old, your friends die, events sweep on and all things are changed. Yet there in your garden or orchard are the birds of your boyhood, the same notes, the same calls.

The swallows, that built so far out of your reach beneath the eaves of your father's barn, the same ones now chatter beneath the eaves of your barn. The warblers and shy wood-birds you pursued with such glee ever so many moons ago, no marks of change cling to them; the whistle of the quail, the strong piercing note of the meadow-lark—how these sounds ignore the years, and strike on the ear with the melody of that spring-time when the world was young.

A Bird Medley.

JOHN BURROUGHS.

Then, all at once, the land laughed into bloom.

ALFRED AUSTIN.

March Flowers

Wild Cyclamen or Shooting-Stars. Common on hillsides.
Brodiaea. Very abundant on sunny hillslopes.
California Lilac. In bosky thickets.
Fuchsia-flowered Gooseberry. Steep cañon-sides.
Ferns. Giving beauty and grace to cañons and hills.
Sun Cups or Golden Eggs. On low slopes.
Bush Lupine. Abundant on cañon-sides.
Calendrinia. Low hillsides.
Filaree. Common carpet of roadside, pasture, orchard and vacant lot.
Eschscholtzia or California Poppy. The golden glory of field and wayside.

> Thy satin vesture richer is than looms
> Of Orient weave for raiment of her kings.

The Eschscholtzia.

INA COOLBRITH.

April Birds

For, lo, the winter is past, the rain is over and gone; the flowers appear on the earth; the time of the singing of birds is come.

SONG OF SOLOMON.

Western House Wren. Very common.
Plain-crested Titmouse. Very common among the live-oaks.
California Bush Tit (Tomtit). Abundant. An early nester.
California Purple Finch. Rather rare.
Black Pewee (Black-headed Flycatcher). Very common.
Bullock's Oriole. Tolerably common.
Red-winged Blackbird. Locally distributed.
Green-backed Goldfinch (Wild Canary). With us all the year round.
Rufous Hummer. A radiant visitor from Central America.

> And here the wild birds sing,
> And there the wild flowers blow;
> My heart—'tis on the wing,
> I know not where 'twill go.

JOHN VANCE CHENEY.

April Flowers

Yellow Pansy. Among the scattered oaks on Boswell's ranche, and at Point Isabel.
Blue-eyed Grass or Nigger Babies. Thick in moist pastures.
Nemophila. In Strawberry Cañon, also on trail to Wild Cat Cañon from North Berkeley.
Wild Oats. An home from hilltop to the bay shore.
Pepper Grass. Moist waysides.
Yellow Mustard. Luxuriant on plain and meadow.
Buttercups. Abundant everywhere.

The flowering of the buttercups is always a great, and I may truly say, a religious event in any year.
The Buttercup.

JAMES RUSSELL LOWELL.

Oh, for the time
Of the mustard's prime;
  For the shifting haze
  Of its yellow maze;
For the airy toss
Of its yellow gloss;
  For the amber lights
  Along the heights
Of the verdurous April ways.

ANNA CATHERINE MARKHAM.

May Birds

Western Flycatcher. Common, nesting in mossy banks.
Warbling Vireo. Common summer resident.
Summer Warbler. Less common of late.
Rufous crowned Sparrow. Fairly common in the hills.
Western Savannah Sparrow. In open fields.
Lazuli Bunting. Common summer resident.
Western Lark Finch. Common summer resident.
Russet-backed Thrush. Abundant. A peerless songster.

>   All the notes of the forest-throng,
>   Flute, reed and string are in his song;
>   Never a fear knows he, nor wrong,
>   Nor a doubt of anything.

The Thrush.

EDWARD ROWLAND SILL.

>   That's the wise thrush; he sings each song twice over
>   Lest you should think he never could recapture
>   The first, fine, careless rapture.

ROBERT BROWNING.

> The voice of one who goes before to make
> The paths of June more beautiful, is thine,
> Sweet May.
>
> Helen Hunt Jackson.

May Flowers

Cream Cups.   Moist hillside fields above North Berkeley towards Grizzly Peak and Wild Cat Cañon.
Fritillaria.   Rich mould of wooded cañons.
Columbine.   In secluded glens, especially in Sir Dagonet's Glen back of Institute, and in Woolsey Cañon.
Tidy Tips, or Yellow Daisies.   Brightening meadow and plain.
Calochortus.   At Boswell's, undisturbed by cultivation.

Fancy the waving, pulsing melody of the vast flower congregations flowing from myriad voices of tuned petal and pistil and heaps of sculptured pollen.

John Muir.

June Birds

Road Runner. Rather rare among the hills.
Rock Wren. Not uncommon in the hills.
Ashy-throated Kingbird. Rather rare.
Lawrence's Goldfinch. Rare.
Black-headed Grosbeak. Common summer resident.
Samuel's Song Sparrow. Very common resident.
Wren Tit. A faithful singer.
Anna's Hummer (Humming Bird). Very common resident.
Allen's Hummer. Not uncommon in summer.

> The least of birds, a jeweled sprite,
> With burnished throat and needle bill,
> Wags his head in the golden light,
> Till it flashes, and dulls, and flashes bright,
> Cheeping his microscopic song.

Field Notes.

EDWARD ROWLAND SILL.

Heart of the Summer is Heart of the Year.
      Mrs. A. D. T. Whitney.

June Flowers

Clarkia. Sunny hillsides. Road to Fish Ranche.
Blue Gilia. Makes patches of color in the fields.
Sunflower. On open plains and hillsides.
Evening Primrose. Exposed places and by roadside.
Indian Pink. Illumines roadsides and borders of thickets.
Collinsia. In shade of oaks and other trees.
Owl's Clover. West Berkeley fields.
Wild Rose. Widely distributed. Blossoms indefatigably early and late.

  As slight a thing as a rose may be
   A stepping stone
  Whereby some soul may step from earth
   To love's high throne.
A Rose.
        Clarence Urmy.

  So sweet, so sweet the roses in their blowing,
  So sweet the lilies are, so fair to see :
  So blithe and gay the humming bird a-going
  From flower to flower, a-hunting with the bee.
        Norah Perry.

July Birds Western Wood Pewee. Common in the woods.
Russet-backed Thrush. Nesting.
Bullock's Oriole. In song.
Black-headed Grosbeak. Singing.
Green-backed Goldfinch. Abundant.
Barn Swallow. Nesting under the eaves of barns.
Cliff Swallow. Nesting.
House Finch (Linnet). Very abundant resident.

> The Power that built the starry dome on high,
> And poised th' inverted rafters of the sky,
> Teaches the linnet with unconscious breast
> To round the inverted heaven of her nest.
>
> ANONYMOUS.

> The shadow of a bird
> On the shadow of a bough,
> Sweet and clear his song is heard;
> "Seek me now, I seek thee now."
> The bird swings out of reach in the swaying tree,
> But his shadow on the garden walk below belongs to me.
>
> EDWARD ROWLAND SILL.

July Flowers

> Through the open door
> A drowsy smell of flowers—gray heliotrope,
> And sweet white clover, and shy mignonette—
> Comes faintly in, and silent chorus lends
> To the pervading symphony of peace.
>
> JOHN G. WHITTIER.

> There are crowds who trample a flower into the dust, without once thinking that they have one of the sweetest thoughts of God under their feet.
>
> J. G. HOLLAND.

> Flowers themselves, whate'er their hue,
> With all their fragrance, all their glistening,
> Call to the heart for inward listening.
>
> WILLIAM WORDSWORTH.

Tarweed. Exasperatingly abundant in the eyes of cross-country walkers.

Yerba Buena. Fringing Strawberry Creek.

Common Monkey Flower. Low moist places in ditches and streambeds.

Godetia. Hillsides, especially toward Claremont.

Wild Honeysuckle. Climbing into trees along Strawberry Creek.

August Birds
Western Chipping Sparrow. Still occasionally trilling its spring song.
Western Lark Finch. In flocks among the fields.
Lazuli Bunting. A beautiful fleck of blue in the thickets.
Plain crested Titmouse. The Quaker of the oak groves.
California Jay. Abundant and noisy.
California Bush Tit. Busy little bands among the live oaks.
Red-shafted Flicker. Always in evidence among the hills.
Western Screech Owl. Its sweet call heard at night.

James Russell Lowell, whose wont it is to see and hear the thing commonly overlooked, regards the cry of this owl, (The Screech Owl,) as one of the sweetest sounds in Nature.

Wood Notes Wild.

SIMON PEASE CHENEY.

The last hour of light touches the birds as it touches us. When they sing in the morning, it is with the happiness of the earth; but as the shadows fall strangely about them, and the helplessness of the night comes on, their voices seem to be lifted up like the loftiest poetry of the human spirit, with sympathy for realities and mysteries past all understanding.

A Kentucky Cardinal.

JAMES LANE ALLEN.

Zauschneria.   Hillsides, mostly in rocky places.
Clematis.   Climbing over shrubs on the cañon-walls.
Twin-Berry.   Tenant of stream-banks and bottoms.
Pimpernel or Poor Man's Weather-glass.   Waste places.
Ripening Grasses, whispering the brown earth's secrets.
Succory.   On low fields stretching to the bay.

    Consider what we owe to the meadow-grasses; with their feathery, or downy seed-vessels, mingling quaint brown punctuation with the bloom of the nearer fields; and casting a gossamered grayness and softness of plumy mist along their surfaces far away.

JOHN RUSKIN.

    In the fields the tall-stemmed blue succory lights one or two blossoms in its chandelier; it is thrifty, and means to have its lamps last, not burn out all at once.

The Seasons.

OLIVER WENDELL HOLMES.

September Birds Red-breasted Nuthatch. An autumn and winter visitor.
Gairdners's Woodpecker. Occasionally found during autumn and winter.
House Finch. Old and young in flocks.
Blue-fronted Jay. Occasional visitor.
Pileolated Warbler. A beautiful visitant during autumn and winter.
Lutescent Warbler. Singing in the cañons.
Green-backed Goldfinch. In flocks among the tarweed.
Meadow Lark. Revives its sweet spring song.

> Oh, for the tryst
> Of the lark in the mist;
>   For the fleeting flash
>   Of his breast's gold plash;
> For the thin fused gold
> Of his song retold,
>   Like a flute's uplift
>   Through the silent rift
> Of an orchestra's dying clash.
>
> ANNA CATHERINE MARKHAM.

Song of the Meadow Lark.

From Wood Notes Wild.

(By permission of Lee and Shepard.)

September Flowers

Asters and Golden Rod. Corners of fields, dry stream-banks and hillsides.
Mallows. Vacant lots.
Thimble Berry. Everywhere in the cañons.
Yellow Sweet Clover. Streets and waste places.
Wild Radish. Everywhere in waste places.
Belated Wild Roses and Poppies.

O sweet wild rose! O strong south wind!
The sunny roadside asks no reasons
Why we such secret summer find,
Forgetting calendars and seasons.

A Wild Rose in September.

HELEN HUNT.

I know the lands are lit
With all the autumn blaze of Golden Rod;
And everywhere the Purple Asters nod
And bend and wave and flit.

HELEN HUNT JACKSON.

October Birds
Arctic Blue-bird. Occasional in flocks during autumn and winter.
Western Golden-crowned Kinglet. A lovely waif from the north-land.
Ruby-crowned Kinglet.
Townsend's Sparrow.
Dwarf Hermit Thrush.
Oregon Junco. A sprightly little winter visitor.
Tule Wren. Common in marshes on the bay shore.
Maryland Yellowthroat. Common in marshes.
Streaked horned Lark. In open fields near the bay.

> These are the days when birds come back,
> A very few, a bird or two,
> To take a backward look.
>
> EMILY DICKINSON.

> A host of poppies, a flight of swallows;
> A flurry of rain, and a wind that follows
> Shepherds the leaves in the sheltered hollows,
> For the forest is shaken and thinned.
>
> EDWIN MARKHAM.

These are the days when skies put on
The old, old sophistries of June,—
A blue and gold mistake.
Till ranks of seeds their witness bear,
And softly through the altered air
Hurries a timid leaf!

EMILY DICKINSON.

Sand-Verbena. West Berkeley.

Blue Curls. Dry fields.

Wax Berry. Near Summit reservoir, and North Berkeley stone-quarry.

Rose-hips and Blackberry vines. Color-bearers along the sides of the creeks.

These few dear Autumn flowers!
More beautiful they are
Than all that went before,
Because they are the last
Of all the Summer's store.

ANONYMOUS.

November Birds American Pipit. Abundant in flocks in open fields.
Oregon Junco (Snow Bird).
Lincoln's Finch. Fairly common in winter.
Say's Pewee. Moderately common winter resident.
Red-breasted Sapsucker. Rather rare winter visitant.
Harris's Woodpecker. Fairly common in winter.
Varied Robin. A shy, solitary, but common winter visitant.

> In the sculptured woodland's leafless aisles,
> The robin chants the vespers of the year.
>
> ALFRED AUSTIN.

All great forms, inanimate or alive, in time, in space, or in mind, are His shadows: all voices, language, music, the inspired word, the sounds and breathings of nature are His echoes.

MOZOOMDAR.

Shrubby Monkey-Flower. Steep south hillsides.
Solanum or Nightshade. Strawberry Cañon.
Coffee Berry. Cañons, and borders of thickets in the higher hills.

There is no glory in star or blossom
Till looked upon by a loving eye.
      WILLIAM CULLEN BRYANT.

There's beauty waiting to be born,
And harmony that makes no sound.
      MRS. A. D. T. WHITNEY.

Wingèd clouds soar here and there
Dark with the rain new buds are dreaming of.
      PERCY BYSSHE SHELLEY.

December Birds

Lewis's Woodpecker. An occasional winter visitant.
Hutton's Vireo. Fairly common during the winter.
Oregon Towhee (Catbird).
Audubon's Warbler. A common winter resident.
Townsend's Sparrow. Solitary, scratching among the leaves.
Gambel's White-crowned Sparrow. One of the few birds that sing during the winter.
Golden-crowned Sparrow. In song.
Samuel's Song Sparrow. Sings at times during the winter.

The sparrows are all meek and lowly birds. They are of the grass, the fences, the low bushes, the weedy wayside places. Nature has denied them all brilliant tints, but she has given them sweet and musical voices. Theirs are the quaint and simple lullaby songs of childhood.

JOHN BURROUGHS.

Gently and clear the sparrow sings
While twilight steals across the sea,
And still and bright the evening star
Twinkles above the golden bar
That in the west lies quietly.

CELIA THAXTER.

If Winter comes, can Spring be far behind?
PERCY BYSSHE SHELLEY.

December Flowers

Toyon or California Holly. University grounds and Cañon.
Mistletoe. Wild Cat Creek.
Laurel. Along Strawberry Creek, and climbs in dwarf form to top of Grizzly.

Can this be Christmas—sweet as May,
With drowsy sun and dreamy air,
And new grass pointing out the way
For flowers to follow, everywhere?

EDWARD ROWLAND SILL.

Before beginning, and without an end,
As space eternal and as surety sure,
Is fixed a power divine which moves to good;
In dark soil and the silence of the seeds
The robe of Spring it weaves.

The Light of Asia.

EDWIN ARNOLD.

OD wills that, in a ring,
His blessings shall be sent
From living thing to thing,
And nowhere stayed nor spent.

JOHN W. CHADWICK.

ONE THOUSAND COPIES OF THIS BOOK HAVE BEEN PRINTED ON STRATHMORE PAPER, FROM THE TYPE, AND TYPE DISTRIBUTED, IN THE MONTHS OF OCTOBER AND NOVEMBER, A. D. MDCCCXCVIII, IN SAN FRANCISCO AT THE SHOP OF THE STANLEY-TAYLOR COMPANY.

www.ingramcontent.com/pod-product-compliance
Lightning Source LLC
Chambersburg PA
CBHW030052170426
43197CB00010B/1490